Contents

Teacher Notes

This guidebook, for use with base-ten blocks, provides hands-on, engaging, and fun discovery lessons. The lessons provide activities to develop, practice and reinforce learning. Utilizing the manipulatives allow students to explore base ten through building numbers, comparing, rounding, adding, subtracting, multiplying, and dividing. Each chapter begins with a letter to parents regarding NCTM Standards and classroom goals for each unit. Family activity suggestions are also included for reinforcement of concepts at home. Problem-solving strategies, as well as implicit and explicit journal/discussion questions allow the teacher to check for ongoing student comprehension of the topic.

CHAPTER 1 Thinking About Math:
Representing and Comparing Whole Numbers

Dear Parents,

Students must gain an understanding of how numbers relate to one another. They need to understand the relationships between ones, tens, hundreds, and thousands. Students also need to be able to recognize that patterns occur when identifying place value.

Through this unit, students will use tools that will build an understanding of place value. Students will be using a set of manipulatives called base-ten blocks. In order for students to gain a deeper understanding of place value, they will be provided many opportunities to explore with these manipulatives. Base-ten blocks will assist in providing a foundation for understanding how to compare and compute numbers. Students will also be given opportunities to reflect about their thinking through writing in journals and oral discussion.

Please consider helping to make deeper connections for your child by completing some of the following activities together. When home and school work together to use the same math vocabulary and provide similar math experiences, the understanding is so much richer.

Sincerely,

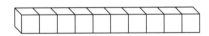

School and Home Connections

❑ Look for numbers greater than one hundred in the newspaper. Cut them out and make a number collage. Write about how the number was used in the article or advertisement.

❑ Find two items in the house that you estimate would weigh the same. Discuss the items that you find.

❑ Count loose change and put it in groups equaling one dollar.

❑ Create money sorting boxes. Make one "hundreds" box for dollars, a "tens" box for dimes, and a "ones" box for pennies.

❑ See how many different combinations of numbers you can create whose sum equal one hundred.

MATERIALS

Math journals

Overhead projector

Base-ten blocks

Number cards
(Appendix G)

KEY WORDS

Standard form

Expanded notation

LESSON 1 Representing Numbers

NCTM Standard: Understanding Numbers
To demonstrate knowledge and use of numbers and their
representatives in a broad range of settings.

GOAL

Students will write and identify numbers in a variety of forms including expanded notation and standard form. Students will represent numbers in a variety of ways using manipulatives including money.

DEVELOPMENTAL INSTRUCTION

1. Display the following journal question to pre-assess student understanding of number identification. Watch to ensure that all students are able to write the number correctly in standard form.

Journal Question: Numbers may be represented many different ways. One way we can show them is through money. The number 21 can be shown by drawing two dimes and one penny. Write the number 137. How would you draw this number using dollars, dimes, and pennies?

2. In small groups of 3-4, have students share their journal responses. Choose one group to share their thinking with the class.

3. Model for students that numbers may be represented in many different ways by displaying base-ten blocks. Ones are represented by units. Tens are represented by rods. Hundreds are represented by flats. Thousands are represented by cubes. Give different examples of how to write what the blocks represent in standard form. For example, place on the overhead 5 flats, 3 rods, and 7 units. Write 537 on the overhead. Tell the students this is how you write in standard form.

4. Provide number cards (Appendix G) for students to practice writing in standard form several times using a variety of numbers from 100-999.

5. Give examples of how to write in expanded notation. For example, place on the overhead 4 flats, 2 rods, and 2 units. Write 400+20+2 on the overhead. Tell the students this is how you write in expanded notation.

6. Provide numbers for students to practice writing in expanded notation several times using a variety of numbers from 100-999.

Stamp It!

1. Place students in pairs.

2. Distribute materials to each pair of students.

3. Place 2 flats, 3 rods, and 7 units on the overhead. Discuss each piece with students to ensure that they all understand what each piece represents. Each rod represents 10 units and each flat represents 100 units. Ask, "What does the total amount represent?" Model illustrating 237 using base-ten stamps instead of blocks.

4. Students are now ready for the "Stamp It" activity worksheet.

Extension

1. Place students in pairs.

2. Cut out the Base-Ten Cards (Appendix F).

3. The cards are shuffled and placed facedown in one pile.

4. Each player chooses 3 cards from the pile and arranges the Base-Ten Block Cards in order from greatest to least. Players then say the number the cards represent in either standard form or expanded notation. Cards are then placed at the bottom of the pile.

5. The player with the higher number wins that round.

6. Play continues for five rounds.

MATERIALS

Base-ten blocks

Stamps (one set for each pair)

Overhead projector

Pencil for each student

Activity 1 worksheet (one for each pair)

MATERIALS

Two sets Base-Ten Cards (Appendix F)

Scissors for each pair

Stamp It!

For each number complete the chart by writing the expanded form. Then use base-ten stamps to illustrate the number. The first one has been completed for you as an example.

Standard Form	Expanded Notation	Stamp It!
164	100 + 60 + 4	
421	400 + 20 + 1	
530	500 + 30 + 0 = 530 =	
279	200 + 70 + 9 = 279	
347	300 + 40 + 7 = 347	

Show Me the Ways

1. Students will work independently.

2. Remind students that numbers may be represented in many different forms (i.e. standard, expanded, notation, or models). Observe ways they are able to name.

3. Tell the students that they are going to show you some of the different forms they have learned by completing Activity 2.

4. Distribute Activity 2 worksheets.

5. When completed, have students share the different ideas they discovered about representing numbers.

Extension

Place students in pairs. Have students cut out the Number Cards (Appendix G) and place them in a pile facedown. Each student chooses four cards and writes the number in their math journal. Students should then write the number in expanded notation. Once the students have created five numbers, they should order the numbers from least to greatest in standard form, in their journals.

MATERIALS

Activity 2 worksheet for each student

Pencil for each student

MATERIALS

Number Cards (Appendix G)

Math journal for each student

Pencil for each student

Show Me the Ways

Use expanded notation, money stamps, or base-ten stamps to show another way of representing the standard form.

1.

43

2.

181

3.

603

4.

1,027

5.

159

6.

262

Base-Ten Rummy

MATERIALS

Base-Ten Rummy cards for each group (Appendices A-D)

Scissors for each group

1. Students will work in groups of 2-4 to play Base-Ten Rummy.

2. Cut out the Base-Ten Rummy cards.

3. The cards are shuffled and placed facedown in front of the players. One player is chosen to be the dealer and distributes four cards to each player. The dealer then turns over the top "Rummy" card to start a discard pile.

4. The player to the left of the dealer goes first. The object of the game is to get all of the cards in your hand to be equal representations of the same number. (i.e. 45, 40+5, 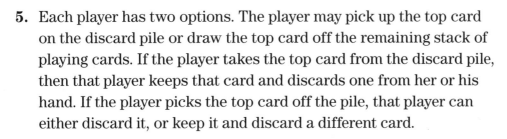 , and $0.45)

5. Each player has two options. The player may pick up the top card on the discard pile or draw the top card off the remaining stack of playing cards. If the player takes the top card from the discard pile, then that player keeps that card and discards one from her or his hand. If the player picks the top card off the pile, that player can either discard it, or keep it and discard a different card.

6. Play continues, in the same direction, until one of the players wins the game.

Hide and Seek – A Game to Review Place Value

DIRECTIONS

1. Players determine numbers by rolling one, two, three, and four number cubes, respectively. One player writes the number rolled in each of the boxes. Both players use the same numbers for each game.

2. Each player then decides where each of the four numbers will be located on their grid and secretly writes the number, either horizontally or vertically, in adjacent squares.

3. The first player starts the game by calling out a square's coordinates (such as "F5"). The second player announces whether the F5 square is a "hit" or a "miss." If the first player calls out the square in which a digit is hidden, then the second player indicates a "hit." If the first player calls out a square in which no digit is hidden, the second player indicates a "miss."

4. If it is a "hit", the first player colors the F5 square red on the blank grid, and the second player puts an "X" through the number on his grid. It is a "miss", the first player simply colors the F5 square blue.

5. A player must announce when a number has been completely colored.

6. Play alternates between players. The winner is the first player to find and completely color in all four of the opponent's numbers.

Hide and Seek Numbers

Ones ☐

Tens ☐ ☐

Hundreds ☐ ☐ ☐

Thousands ☐ ☐ ☐ ☐

| Blue Miss | | Red Hit |

Names _____

_____

Date _____

Hide and Seek

	1	2	3	4	5	6	7	8	9	10	11	12
A												
B												
C												
D												
E												
F												
G												
H												
I												
J												
K												
L												

MATERIALS

Math Journal for each student

Number Cards
(Appendix G)

Base-Ten Cards
(Appendix F)

LESSON 2 Comparing Numbers

NCTM Standard: Understanding Numbers
To understand the place-value structure of the base-ten number system and be able to represent and compare whole numbers and decimals.

GOAL

Students will compare whole numbers.

DEVELOPMENTAL INSTRUCTION

1. Display the following journal question to pre-assess student understanding of comparing numbers.

Journal Question: How would you tell a friend how to decide whether a number is greater or less than another number?

2. Have students share with a friend how they determined a greater or lesser number. Choose one pair to share their thinking with the class.

3. Using number cards from Appendix G, create the number 4,671 and 4,761. Give each group of three to four students 8 cubes, 13 flats, 13 rods and 2 units cut from Appendix F.

4. Have students create the numbers modeled on the overhead. In their math journals, students should record the two numbers in standard form.

5. Ask students to explain the meaning of each digit orally or continue the explanation in their math journals. Repeat by creating two new numbers for students to compare. (i.e. 3,346 and 2,346)

6. Ask for volunteers to create new examples for the class to try.

7. When students have a good understanding of how to compare numbers, begin Activity 5-Can of Worms.

Can of Worms

1. Place students in pairs.

2. Give each pair an Activity Mat (Appendix E), a number cube, and Spinner (Appendix H), so they may play "Can of Worms".

3. Player A goes first and tosses the number cube. Player A is blue.

4. Player A records the number in the ones column and tosses the cube again, this time recording the number in the tens column. The third toss is written in the hundreds column.

5. Player A then creates this number with base-ten blocks.

6. Player B is red. Player B tosses the number cube and repeats steps 4-5.

7. Player A spins the spinner. If the spinner lands on "Less Than", the player with the lesser number wins the round and places a "worm" (a piece of his or her string) into the container. If the spinner lands on "Greater Than", the player with the greater number wins the round and places a "worm" into the container.

8. Play five rounds or for a predetermined amount of time.

9. The player with the most of his or her color worms in the container wins the game.

MATERIALS

Base-ten blocks for each pair

Number cube for each pair

Activity mat for each pair (Appendix E)

Spinner for each pair (Appendix H)

Container for each pair

Five pieces of 6" blue string for each pair

Five pieces of 6" red string for each pair

Build a Number

1. Students will work independently for this activity.

2. Distribute materials, one worksheet and number cube per student.

3. Explain to students the connection between units and ones, rods and tens, flats and hundreds, and cubes and thousands.

4. Students will numerically represent numbers in the thousands. For example, Jo built the number 3,341. Students will create the number using their base-ten blocks.

5. Students will complete the "Build a Number" chart through Hector's row.

6. Discuss with students the values created thus far, then pose the journal question.

Journal Question: After creating numbers for the first four children, can you think of a strategy for creating larger numbers with the number cube? How about smaller amounts? Apply this strategy to the remaining children.

7. Students complete the remaining rows of the "Build a Number" worksheet.

Name _____

Date _____

Build a Number

Toss a number cube four times to create a number in the thousands. The first toss represents the ones place. The second toss represents the tens place. The third toss becomes the hundreds place and fourth toss is the thousands. Record each number in the correct place value column. Write the whole number next to the person's name. The first one has been done for you.

Build a Number	Toss 4 Thousands	Toss 3 Hundreds	Toss 2 Tens	Toss 1 Ones
Jo 3,341	3	3	4	1
Laura				
Bryan				
Hector				
Anna				
Kaitlyn				
Jamey				
Jackson				

Name _____

Date _____

Chapter 1 Assessment: Whole Number Wrap-Up

⟵――――――――――――――――――――――――――――――⟶

25 30 35 40 45 50 55 60 65 70 75 80 85 90 95 100 105 110 115 120 125 130 135 140 145 150

Use < or > to complete this problem.

1. 53 ◯ 63   **2.** 29 ◯ 31   **3.** 84 ◯ 82   **4.** 107 ◯ 110   **5.** 50 ◯ 60

6. 113 ◯ 123   **7.** 530 ◯ 503   **8.** 992 ◯ 993   **9.** 10 ◯ 100   **10.** 2,003 ◯ 3,002

Order the numbers from least to greatest.

1. 48; 840; 484; 4,008; 804   **2.** 1,001; 1,010; 10; 110; 11

_____   _____

3. 3,229; 229; 92; 2,929   **4.** 630; 3,006; 360; 3,363

_____   _____

Think About It! How can a number line help you compare numbers?

_____

_____

_____

Chapter 2 Thinking About Math:
Understanding Decimals

Dear Parents,

Students must gain an understanding of the relationship between whole numbers and parts of whole numbers (decimals). The relationship between these numbers is sometimes confusing and frustrating for students to grasp. Throughout this unit, students will use a variety of manipulatives that will help build an understanding of decimal place value. One tool students will use is base-ten blocks. This hands-on tool, in conjunction with a variety of decimal games and activities, is an effective method for exploring and gaining mastery of decimal place value.

Please consider completing a few of these fun reinforcing activities together with your child. Thank you for your continued help. Valuable experiences are available to our children and a deeper understanding arises when we can make that practical connection of using math in our daily lives.

Sincerely,

School and Home Connections

❑ Purchase or look around your home for a 100-piece puzzle. Count the pieces and explain that all the pieces together make one whole number. 100 pieces = 1 puzzle. Experiment by taking 5 pieces from the puzzle. What do you have left? (.95) Continue this exercise by removing a variety of numbers. Discuss and record your findings.

❑ Look in the sports section of the newspaper. Record and compare batting averages of your top three favorite players. Look for averages in other sports as well.

❑ Discuss ways in which decimals are used in the business section of the newspaper.

❑ Use grocery advertisements to compare prices. Find which store offers the best value for your favorite kind of fruit.

MATERIALS

100-piece puzzles—one for each group of four students

Math journal for each student

Pencil for each student

Overhead projector

Appendix I (Transparency)

KEY WORDS

Fractions

Percents

Decimals

LESSON 1 Puzzling Decimals

NCTM Standard: Understanding Numbers
To understand the place-value structure of the base-ten number system and be able to represent and compare whole numbers and decimals.

GOAL

Students will recognize and create equivalent forms of commonly used decimals.

In Lesson 3, students will be introduced to the concept of decimals equating fractions and percentages to the hundredths place. The following two lessons and the corresponding activities will lead the students toward this goal.

DEVELOPMENTAL INSTRUCTION

While showing students a 100-piece puzzle box, tell them that when all of the pieces are put together the completed project is one whole picture. Write 100/100=1 on the board. **Journal Question:** How could you mathematically represent the puzzle if fifty pieces remained in the box? List the difference representations. You should see answers such as, ½ and 50%.

1. In groups of four, have the students brainstorm places they have seen decimals used.

2. Create a web on the board, linking similar ideas together.

3. From this web, categorize decimal use in three areas: for counting, for measuring, and to tell order.

4. Keep students in their small groups and distribute 100-piece puzzles. Have students complete their team puzzle, emphasizing that they are each creating one whole.

5. When students have finished, model the removal of ten pieces of this puzzle. Place a transparency of Appendix I on the overhead projector. (Work in increments of ten.) Record findings in three forms (i.e. 90/100 = 90% = .90).

6. Continue direct instruction by telling the students to imagine that the whole puzzle is the same as $1.00. Fractions, percents, and decimals show that numbers can represented in different ways.

7. Begin Activity 8-Money / Decimal Mat.

Money/Decimal Mat

MATERIALS

Money cards
(Appendix J) for
each student

Activity 8 worksheet
for each student.

Scissors for each
student.

1. Remind students that $1.00, like a completed puzzle, represents one whole and is written 1.0.

2. Students will work independently with money cards and activity worksheet 8 to create equivalent numbers.

3. Distribute money cards and worksheets, so that each child has his or her own copy.

4. Students cut apart the money cards and flip them upside down.

5. Each student chooses three cards from the pile in random order and places them in the appropriate section of the Money/Decimal Value Mat (i.e. dollars go in the whole numbers column, dimes in the tenths column, and pennies in the hundredths column).

6. Students record the created number on their mat.

7. Cards are returned back to the pile facedown.

8. Students repeat the process by selecting three more cards. They record this new number in row number two.

9. After creating ten new numbers, students should order them from least to greatest.

Name _____

Date _____

Money/Decimal Mat

Whole Numbers	Tenths	Hundredths
1.		
2.		
3.		
4.		
5.		
6.		
7.		
8.		
9.		
10.		

LESSON 2 Equivalent Decimals

NCTM Standard: Understanding Numbers
To understand the place-value structure of the base-ten number system and be able to represent and compare whole numbers and decimals.

GOAL
Students will gain a deeper understanding of equivalent decimals.

Journal Question: Numbers can be represented in different ways. We will use base-ten blocks to show decimals. Like a puzzle, a flat represents a whole (i.e. 100 pieces=100/100=1). A rod, therefore, represents tenths (.1). What is a unit most likely to represent? How did you come to this idea?

DEVELOPMENTAL INSTRUCTION
Whole numbers are placed in front of a decimal point. After the decimal point the numbers represent pieces of a whole or fractions of the number.

1. Think about the 100-piece puzzle.

2. Place a 100-piece puzzle on the overhead. Ask, "if one piece is removed, how many are left?"

3. Record your thinking in your math journals by showing the answer in decimal form. (.99) Now read your answer aloud. (ninety-nine hundredths)

4. Pretend you have two 100 piece puzzles. What if five pieces were missing from one? How would you write this number? (one and ninety-five hundredths) Remember, when reading the number, the decimal point is read as "and".

5. Continue with different scenarios to see if students are able to record and read decimals.

MATERIALS

Math journal for each student

Pencil for each student

Appendix F or base-ten stamps

Overhead projector

Completed Activity 8 worksheet

KEY WORD

Equivalent

Money Mat and Base-Ten Blocks

1. Have students take out their completed Activity 8 worksheets.

2. Distribute materials (Appendix F or base-ten stamps and pencils).

3. Students will create equivalent (equal) decimal amounts for problems 1-10. (i.e. $2.33=2 flats, 3 rods, 3 units). They should record picture representations in their math journal.

Extension

Use the numbers from problems 1-10 in Activity 8 to add together decimal numbers. Have students add together the numbers from 1 and 2, 3 and 4, 5 and 6, 7 and 8, 9 and 10. Students should have five new decimal numbers and record these in their journals. Then have students draw pictures of base-ten blocks or use the base-ten stamps to represent the sum of each pair of decimals.

LESSON 3 Comparing Decimals

NCTM Standard: Understanding Numbers
To understand the place-value structure of the base-ten number system and be able to represent and compare whole numbers and decimals.

GOAL

Students will compare decimals, fractions, and percents.

Journal Question: People buy different kinds of milk. The fat content is displayed as a percent. What kind of milk does your family buy? Represent this number as a percent, fraction, and decimal (i.e. half and half=50%, ½, .5). Tell students that decimals are used for three reasons: to count, to measure, and to tell order. Ask them to think of places, other than at the grocery store, where decimals might be used for these reasons.

DEVELOPMENTAL INSTRUCTION

1. Distribute Writing Numbers in Three Ways (Activity 10) worksheet to each student.

2. Place base-ten blocks in a bag or bucket. A student closes his or her eyes, reaches in and takes out a combination of flats and rods.

3. Place these manipulatives on the overhead projector.

4. The students' job is to count the tenths and hundredths and color the corresponding amount in space number 1 on the worksheet.

5. Instruct students that there are three different ways to represent each number. A flat represents a whole. Therefore a rod represents 1/10 or 0.1 of a flat because ten rods make a flat. Since there are 100 units in a flat, a unit represents 1/100 or 0.01 of a flat (i.e. two rods and seven units = .27, 27% and 27/100). Choose student volunteers to reach in the bag and make new amounts of numbers. Continue making new numbers and record them in three different ways on Activity 10 worksheet.

MATERIALS

Base-ten blocks (flats and rods only)

Overhead projector

Activity 10 worksheet for each student

Thin markers or color pencils

ACTIVITY 10

Writing Numbers in Three Ways

Decimal	Percent	Fraction
1.		
2.		
3.		
4.		
5.		
6.		
7.		
8.		
9.		
10.		

Memory Match

MATERIALS

Memory Match
cards for each pair
(Appendix K)

Scissors for each pair

1. Distribute one set of Memory Match cards per pair of students. It is best if the cards are photocopied on card stock.

2. Have students cut cards apart and shuffle them.

3. Players place cards facedown in a 4x4 array.

4. Player A turns over four cards. If they are all equivalent, the player keeps the set. If they are not all equivalent, the player turns the cards over to their original places.

5. Player B repeats, trying to find four equivalent numbers.

6. Play continues until all matches have been found. The player with the most matches is the winner.

Activity 12 worksheet for each group

Newspapers, catalogs, almanacs, Guinness Book of World Records, rulers, radios, and other resource materials

Chapter 2 Assessment: Scavenger Hunt

1. Remember decimals can also be represented as fractions and percents. Using Activity 12 worksheet, have students go on a scavenger hunt using resource materials and items in their school to find the use of decimals in their environment. Students should work in teams of two to three members.

2. As students find decimals, percentages, and fractions, they should record the number and how the decimal, percent, or fraction was used in the proper column.

3. After fifteen minutes, have students share their findings in order to generate ideas for other groups.

4. Continue until most resources have been exhausted.

5. Display each teams' results and compare findings. Discuss the most interesting finds.

Extension

Have students take home a copy of the chart and see what other decimals they can add to their collection from home. Tell them to look in the car, at the gas station, the grocery store, hardware store, etc.

Names _____

_____

Date _____

Scavenger Hunt

Decimal	Percent	Fraction
1.		
2.		
3.		
4.		
5.		
6.		
7.		
8.		
9.		
10.		

Chapter 3 Thinking About Math:
Addition and Subtraction of
Two- and Three-Digit Numbers

Dear Parents,

Students must develop fluency in adding and subtracting whole numbers. They must be able to use strategies to estimate the results of whole-number computations and be able to judge the reasonableness of such results.

Throughout this unit, students will use various tools along with kinesthetic activities to build a concrete understanding of the inverse relationship between addition and subtraction. Once again, base-ten blocks will be used to reinforce the students' understanding of the role of place value in addition and subtraction.

Estimating is a vital life skill. During this unit we will explore estimating and rounding together with addition and subtraction. Rounding numbers are approximate and easier to work with for mental computation. Rounding is used to get an answer that is close but not exact.

Sincerely,

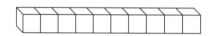

School and Home Connections

❑ Add the number of bananas in a fruit bowl together with the number of oranges. Subtract the bananas. Discuss which fruit is the greater and lesser amount.

❑ Place several paper clips at the end of a piece of paper. Group the paper clips by sliding them along the paper to create many addition and subtraction problems. Add or subtract the first group to or from the second group. Record your findings.

❑ Measure, weigh, or count several items in your home. For example, measure the tallest person and shortest person in your family. Compare these amounts by adding and subtracting your findings. Discuss.

❑ Estimate the amount of milk, butter, peanut butter, and bread your family has used so far this week. Predict when these basic supplies will need to be replaced and estimate the amount of money it will cost to replace them.

LESSON 1 Rounding Numbers

NCTM Standard: Understanding Numbers
To understand place-value structure of the base-ten system by adding two- and three-digit numbers.

GOAL

Students will use rounding to the nearest 10 and 100 to make numbers that are easier to work with for mental computation.

DEVELOPMENTAL INSTRUCTION

1. Instruct students that the purpose of rounding numbers is to create numbers that are easier to work with for mental computation. Rounded numbers are only approximations and they are used to get a close (but not exact) answer.

2. Use Appendix L, a picture of a hill, to demonstrate rounding. The numbers that end in a one, two, three or four, are rounded to the next lower number that ends in zero. Numbers that end in five or greater should be rounded up to the next number that ends in zero. For example, place number 68 at the top of the hill. The teacher should tell the students that they will be practicing to round numbers to the nearest ten.

3. On the right side of the hill write a 70. On the left side of the hill write the number 60. Ask the students to look at the ones place in the number 68. Which way would the number roll towards, the 60 or the 70? By looking at the hill, they should be able to determine that the number would roll towards the 70 because it ends in a number greater than 5.

4. Repeat the activity with the numbers 32, 74, 56, 15, 81 and 47. Distribute copies of Appendix L (the hill) for students in use.

   a. Record the following numbers for rounding to the nearest ten on the overhead: 58, 94, 12 86.

   b. Give students the challenge of rounding the following numbers to the nearest hundred: 631, 424, 785, 638, 912.

   c. Continue the challenge by rounding to the nearest thousand and/or ten-thousand.

MATERIALS

Transparency of Appendix L, and one copy for each student

Overhead projector

Pencil for each student

KEY WORDS

Rounding

Approximations

MATERIALS

Base-Ten Place-Value
Mat (Appendix M),
one for each student

Base-ten blocks
or base-ten cards
(Appendix F)

Activity 13 worksheet
for each student

Round-Up Rodeo

1. Use the Base-Ten Place-Value Mat (Appendix M), base-ten blocks or base-ten cards (Appendix F).

2. Using the Round-Up Rodeo worksheet, model the first problem. Round 37 to the nearest ten.

3. Place 3 rods in the tens column and 7 units in the ones column to show 37.

4. Remind students that the number to the right of the 3 must be 5 or greater than 5, in order to change the 3 to a 4 or rather round up to the next tens. Seven is greater than 5, so replace the 7 units with another rod in the tens place. The new rounded number is 40.

5. Model the first number in the hundreds place. Round 734 to the nearest hundreds.

6. Place 7 flats in the hundreds column, 3 rods in the tens, and 4 units in the ones column to show 734.

7. Look at the number to the right of 7. If that number is 4 or less, than don't change the 7. Take away all the rods and units to create a new rounded number of 700.

8. Have students complete Round-Up Rodeo while working in groups of two.

Round-Up Rodeo

Use the Base-Ten Place-Value Mat and base-ten blocks to help solve these problems.

Round to the nearest ten:

37	_____	55	_____
54	_____	67	_____
78	_____	91	_____
21	_____	12	_____
33	_____	88	_____

Round to the nearest hundred:

734	_____	109	_____
825	_____	667	_____
249	_____	343	_____
550	_____	487	_____
980	_____	111	_____

Round to the nearest thousand:

994	_____	7,894	_____
4,566	_____	3,356	_____
8,823	_____	5,523	_____
2,909	_____	1,775	_____
5,200	_____	6,455	_____

MATERIALS

Transparency of
Appendix N

Transparency of
Appendix O

Overhead projector

Base-ten blocks
(1 flat, 9 rods, and
9 units per student)

KEY WORDS

Rounding

Estimate

Difference

LESSON 2 Estimating Sums and Differences

NCTM Standard: Understanding Numbers
To understand place-value structure of the base-ten system by adding
two- and three-digit numbers.

GOAL

Students will make estimates of sums and differences in order to help
them to better understand a problem. Students will use estimating to
decide whether the sum or difference of a number is probable.

DEVELOPMENTAL INSTRUCTION

1. Create a classroom working definition of the word *estimate*
   by brainstorming as a group words that define its meaning.
   Look for students to incorporate words such as: reasonable,
   probable, practical, predict, and rounding. Use the words students
   brainstormed to create a web (Appendix N). On the edge of the web
   write some reasons why someone would choose to estimate rather
   than get the exact answer. Look for students to give examples such
   as: estimating the grocery bill, estimating the time it will take to get
   ready for school, estimating how much cereal can be eaten in the
   morning, estimating the sum of two numbers, and estimating if there
   is enough money in the bank to buy a new toy.

2. Model how rounding and estimating work together to give an
   approximate answer, by displaying the page from the estimating store
   (Appendix O). Tell the student that you have $1.00 to spend. On the
   transparency price the items from $.05-$.99.

3. Select one item to buy. Round that number to the nearest ten and take
   that many base-ten blocks to represent the rounded number. Estimate
   aloud to the students how much money you have left to spend. For
   example: If an item is priced at $.47, take five rods aside, showing that
   you have spent approximately $.50 and have a remaining $.50. Use the
   words "rounding", "estimate" and "difference" aloud.

4. Model again by selecting two or three different items. Find the sum
   of the estimated total as well as the difference from $1.00. This time,
   have students mimic by using their own blocks.

Estimating Store

1. Continue the developmental lesson by challenging the students with new prices for the items on the Estimating Store page.

2. Distribute materials to students. Have them record their purchases and estimates in their math journal. You may give students a new amount to spend accordingly.

3. Each child should "visit the Estimating Store" for ten different shopping trips to provide repeated practice of rounding and estimating.

4. Proceed to Activity 15 – Estimation Market.

Extension

Journal Question: Pretend it is a family member's birthday. You have $10.00 to spend. What would you purchase? Where would you shop?

MATERIALS

Appendix O, one copy for each student

Base-ten blocks (5 flats, 9 rods, and 9 units per student)

Math journal for each student

Estimation Market

Suppose you have $1.25. Estimate if you have enough money to buy the following items from the market. Write your estimate in the space provided.

	Estimate	**Do you have enough money? (Yes or No)**
1. Two apples, one watermelon	_____	_____
2. Two potted plants and one pansy	_____	_____
3. Three eggplants	_____	_____
4. Five bananas	_____	_____
5. One pumpkin, one bunch of grapes	_____	_____
6. Six bunches of grapes	_____	_____

What would you buy with your money? _____

_____

Journal Question: What did you need to do to the prices to understand which numbers to round up and which numbers to keep the same? How will rounding numbers help when you are adding and subtracting prices?

Scan, Estimate, and Compare

1. Distribute base-ten materials to students.

2. Each student will begin by placing a handful of base-ten blocks (random amount) on the table.

3. Students should quickly scan and estimate how many flats, rod, and units there are on the table. Have students record the estimate in their math journals.

4. Next, have students count the exact number of blocks and record this number in their journals.

5. Finally, have students compare the two amounts and write their discoveries in their journals.

6. Now have students continue the same practice exercise using the Activity 16 worksheet instead of the base-ten blocks.

MATERIALS

Base-ten blocks (flats, rods, and units)

Activity 16 worksheet for each student

Math journal for each student

Pencil for each student

Name _____

Date _____

Scan, Estimate, and Compare

Scan and Estimate	Exact Number	Compare the Numbers	Base-Ten Blocks
1.			
2.			
3.			
4.			

LESSON 3 Addition and Subtraction

NCTM Standard: Understanding Numbers
To understand place-value structure of the base-ten system by adding two- and three-digit numbers.

GOAL

Students will use problem solving and addition to try and race to the number 100.

DEVELOPMENTAL INSTRUCTION

1. Place students into groups of four and give each group Appendix P, which shows a mountain. Tell the students that they will be racing each other to the top of the mountain. At the top of the mountain is the number 100. They will use their number cards to form addition problems. The sum of the two numbers tells the student how many steps he or she is able to climb up the mountain. Students will need to strategize in order to create addition problems that will give them the largest sums.

2. Students cut apart the number cards. The cards are shuffled and placed in a pile in the middle of the group.

3. Students each draw one card to see who begins play. The player who draws the largest number goes first.

4. Those cards are returned to the pile.

5. Player 1 draws four cards and forms two 2-digit numbers. After creating these numbers, the player must add them together to find the sum. The player colors that number of steps up the mountain.

6. Players 2-4 continue the play, coloring the corresponding sum on their individual mountains.

7. Play continues until a player reaches 100. Players can go over 100.

Journal Question: Which numbers were you most anxious to see when you drew cards? Why?

MATERIALS

Crayon or marker for each student

Sheet of paper for each student

Number cards (Appendix G) (two sheets per student)

Climb the Mountain Appendix P (one per student)

KEY WORDS

Addition

Sum

MATERIALS

Meet the Neighbors
mats (Appendix Q)
(one per pair)

Base-ten blocks
or base-ten cards
(Appendix F)

Number cards
(Appendix R)

Math journal for
each student

Meet the Neighbors

Reminder: When 10 units (or ones) are placed together, you have 1 ten. When 10 rods (or tens) are placed together, you have 1 hundred.

Question: How many hundreds would you need to have 1,000? How do you build that number using base-ten blocks?

Procedure: Students work in pairs to add two numbers together.

1. Each pair cuts apart the number card from Appendix R.

2. The cards are shuffled and placed facedown.

3. Player A chooses a card and builds that number on the mat (Appendix Q) by placing flats, rods, and units on top of the correct house.

4. Player B chooses a card, creates that number on the same mat, and makes the correct exchanges (i.e. numbers 147 and 25: Player B removes ten units from the ones and replaces it with a rod which is now placed in the tens house.) Students count the blocks to find the total sum.

5. Continue playing, alternating player A and B, until a total of ten addition problems have been picked from the card pile and solved.

6. Have students record each problem in their math journals.

Moving Day

It is moving day for the Base-Ten family. They are leaving their Hundred's Home. They have 400 items to be moved.

1. Students should place four flats on the Hundred's House. It will be their job to subtract items from 400 until they have moved all items out of the home.

2. Partners cut apart number cards (Appendix S), turn them facedown, shuffle and place them in a pile.

3. Partners decide who is going to be Mover 1 and Mover 2. Mover 1 draws a card and subtracts that number from 400 by removing flats and making exchanges with the tens and ones houses. (i.e. Mover 1 chooses 60. Mover 1 exchanges one flat for ten rods and replaces three flats on the Hundred's Home, four rods on the Ten's Home, and removes six rods from the homes). The player then keeps the card, representing his or her points for that round.

4. Mover 2 now repeats the steps, subtracting from what remains from Mover 1's difference.

5. If a number is too great to subtract, the mover loses a turn, and the card must be returned to the pile. Play continues until no cards remain, or no more subtraction problems are possible.

6. The movers then add their own cards together to determine points. The mover with the higher sum wins.

7. Distribute Activity 18 Worksheet for reinforcement of addition and subtraction skills.

MATERIALS

Meet the Neighbors mats (Appendix Q) (one per pair)

Base-ten blocks or Appendix F (two copies per pair)

Number Cards Moving Day (Appendix S) (one set per pair)

Calculator

Activity 18 worksheet (one per student)

Mystery Numbers

Use base-ten blocks, stamps, or Appendix F. On a separate sheet of paper, draw the units, rods, and flats that you use to help you complete this worksheet. Solve the mystery number x.

1. $x + 5 = 10$

2. $300 - x = 10$

3. $x + 6 = 8$

4. $60 + x = 100$

5. $20 - x = 10$

6. $43 + x = 60$

7. $10 - x = 6$

8. $285 - x = 165$

9. $110 - x = 100$

10. $155 + x = 200$

Name _____

Date _____

Sums and Differences That Balance

What sums and differences can you create to balance each scale?

EXAMPLE: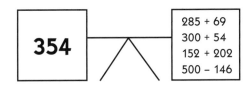

354 | 285 + 69
300 + 54
152 + 202
500 − 146

1.

497

2.

289

3.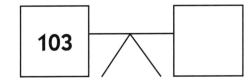

103

4.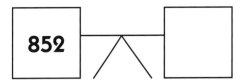

852

5.

511

6.

690

7.

315

8.

702

Saving and Shopping Vacation

The children in the Ruiz Family have been saving their money to spend while on vacation. How much has each child saved? You may use base-ten blocks to solve each problem.

1. Jamal has 2 five-dollar bills, 6 one-dollar bills, 3 quarters, and 2 dimes. $_____

2. Mary has 1 ten-dollar bill, 7 quarters, 3 nickels, and 1 penny. $_____

3. Yolanda has 12 quarters, 12 dimes, 6 nickels, and 23 pennies. $_____

4. Anna has 3 five-dollar bills, 2 one-dollar bills, 4 quarters, and 2 nickels. $_____

5. Who has saved the most money? _____

6. Who has saved the least money? _____

7. At the store, Jamal bought six postcards at $.20 each, six stamps at $.37 each, and one conch shell for $2.50.

   How much did he spend? _____

   How much money does Jamal have left? _____

8. Mary bought one pair of sunglasses for $6.00, some lotion for $1.75, and a postcard for $.20.

   How much did she spend? _____

   How much money does Mary have left? _____

9. Yolanda bought a candy bar for $.65, an inner tube for $.99, and sandals for $1.75.

   How much did she spend? _____

   How much money does Yolanda have left? _____

10. Anna bought a new towel for $10.68, a magazine for $3.59, and a juice drink for $.79.

    How much did she spend? _____

    How much money does Anna have left? _____

Race to Zero

Students will use problem solving and subtraction to try and race each other to the number zero without going below zero. In order to play this game, the students must find the difference of two numbers by solving self-generated subtraction problems.

1. Divide students into pairs. Give each pair a set of number cubes, pencil, two pieces of paper, and a calculator.

2. Tell students to write the number 100 at the top of their sheet of paper and 0 at the bottom.

3. The students must decide who will be player A and who will be player B.

4. Player A tosses the number cubes and creates a two-digit number (i.e. 3 and 2 could be 32 or 23). The player then subtracts that number from 100. The player then checks the answer on a calculator.

5. Player B takes a turn repeating step four.

6. Play continues alternating between Players A and B.

7. The winner is the player who came closest to zero without going below zero.

8. Pose the following **Journal Question:** What strategies did you use to try and win the Race to Zero? What would you do differently if you played the game again?

9. Have the students share their responses with their partner.

10. Have students repeat the game with a new partner.

11. At the conclusion, partners should once again share their thinking. What kinds of new strategies did they try?

MATERIALS

Pencil for each student

Sheet of paper for each student

Number cubes (two per pair)

Calculator (one for each group)

KEY WORDS

Subtraction

Difference

MATERIALS

Overhead projector

Spinners
(Appendix U)

Base-ten blocks
(or Appendix F)

One-centimeter
grid transparency
(Appendix T)

KEY WORD

Perimeter

LESSON 4 Perimeter

NCTM Standard: Use visualization, spatial reasoning, and geometric modeling to solve problems.

GOAL

Students will use base-ten blocks to understand perimeter.

DEVELOPMENTAL INSTRUCTION

Display the following journal question to assess what students know about perimeter.

Journal Question: What is perimeter? Using base-ten rods and units name some ways you can find the perimeter of your math book.

1. Discuss the students' definition of perimeter. Define perimeter as the distance around the border of a two-dimensional figure. To find the perimeter of a square you add together the 2 lengths with the 2 widths.

2. Place a one-centimeter grid transparency (Appendix T) on the overhead. Inform students that one base-ten unit is equivalent to one centimeter.

3. Use spinners (Appendix U) to select how many of each: flats, rods, and units to place on top of the grid paper (i.e. first spinner lands on rods and second spinner lands on 3, you will need 3 rods).

4. Create shapes using the blocks and find the perimeter of the shapes you create.

5. Give the formula for perimeter. $P = 2L + 2W$.

6. Have student volunteers take turns spinning the spinners to create shapes and find the shapes perimeter.

7. Complete several examples. Then distribute Activity 22 – Perimeter Pals.

Perimeter Pals

What is the perimeter of this pal? _____

Create one of your own perimeter pals using base-ten units. Draw it on the grid paper below.
What is the perimeter of your pal?

Perimeter Packages

Use base-ten units to measure the perimeter of each package. (Remember 1 unit = 1 cm.)

1.

   P = _____

2.

   P = _____

3.

   P = _____

4.

   P = _____

5.

   P = _____

6.

   P = _____

7. Draw another package. Find the perimeter.

P = _____

Perimeter Hunt

Using rods and units, measure the perimeter of the following objects around your classroom. Answers should be in centimeters. (Remember 1 unit = 1 cm.) P = 2L + 2W

1. Student's desk top _____

2. Teacher's desk top _____

3. Mouse pad _____

4. Notebook paper _____

5. Seat of a student chair _____

6. Computer screen _____

7. Student's science book _____

8. Tile on the floor _____

9. Library book _____

10. Student's math book _____

MATERIALS

Number cube for each student

Activity 25 worksheet for each student

Pencil for each student

Chapter Three Assessment: Digits Family Collection

The Digits Family collects stamps. Help them figure out how their collection would increase if they added stamps in increments of one, ten, one hundred, and one thousand. Also see what would happen if they sold some stamps and decreased their collection by subtracting one, ten, one hundred, and one thousand stamps.

1. Toss the number cube four times. The first toss represents ones, the second toss represents tens, the third toss is hundreds, and the fourth toss represents thousands. This is the starting number of stamps in the collection.

2. Record the number in each box of the Starting Number row. Then record the number of stamps in the collection by adding and subtracting stamps according to the directions in each column. The sum should be written in the top row and the difference should be written in the bottom row.

Name _____

Date _____

Chapter 3 Assessment: Digits Family Collection

Sum				
+	**Add 1000**	**Add 100**	**Add 10**	**Add 1**
Starting Number				
–	**Subtract 1000**	**Subtract 100**	**Subtract 10**	**Subtract 1**
Difference				

Sum				
+	**Add 1000**	**Add 100**	**Add 10**	**Add 1**
Starting Number				
–	**Subtract 1000**	**Subtract 100**	**Subtract 10**	**Subtract 1**
Difference				

Sum				
+	**Add 1000**	**Add 100**	**Add 10 lbs**	**Add 1**
Starting Number				
–	**Subtract 1000**	**Subtract 100**	**Subtract 10**	**Subtract 1**
Difference				

CHAPTER 4 Thinking About Math:
Multiplication and Division of Whole Numbers

Dear Parents,

Students must gain an understanding of the relationship between multiplication and division. They also must be able to recognize what the numbers in each situation represent. Modeling multiplication and division problems helps extend a student's understanding of the relationship between the two operations. In this chapter, we will use base-ten blocks to further explore the inverse relationships by using pictures, diagrams, and concrete materials. These materials will help the students conceptualize what the terms *factor*, *product*, *quotient*, *divisor*, and *remainder* represent. Below are some ideas to try at home together.

Sincerely,

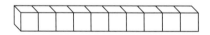

School and Home Connections

❑ Count the number of pens, pencils, or crayons you find in your home. Place them in equal groups. Write division and multiplication sentences. (i.e. How many equal ways can 20 crayons be divided?) Note: Some numbers will have remainders.

❑ Place pocket change on the table and count it. Think of ways you can equally group this amount (divide). Record your findings. (i.e. If you have $1.20 in change. How can it be divided equally among 4 members of your family?)

❑ Find groups of items (fingers on hands, spider legs, pet's legs, bicycle wheels, tricycle wheels, etc.) and create multiplication and division problems. Record findings. (i.e. How many legs do 4 spiders have in all? 4 x 8 = 32. 32 legs total divided by 8 legs on each spider equals 4 spiders.)

LESSON 1 Multiplication

NCTM Standard: Multiplication
Develop fluency in multiplying whole numbers.

GOAL

Students will multiply single-digit whole numbers through multiples of six.

DEVELOPMENTAL INSTRUCTION

Journal Question: Write your own definition of *multiplication* (no dictionaries allowed). Why is multiplication a necessary operation in mathematics?

1. Select a student to orally read the dictionary's meaning of multiplication. Compare it to the definitions generated by the class. Discuss why multiplication is a necessary operation in mathematics.

2. Put students into small groups of 2-3.

3. Place counters in some or all of the six sections of the egg cartons. For example, place two counters in each of two sections. How many counters are there in all? Count by twos out loud. Then write the equation 2 x 2 = 4.

4. Try another problem by placing 5 counters in 3 egg sections, count out loud by fives. Write the equation 5 x 3 = 15.

5. Toss two number cubes. Suppose they land on 3 and 4. Explain that the numbers represent either 3 groups of 4 or 4 groups of 3.

6. Have students toss the number cubes and place the corresponding number of counters in each egg carton, recording the multiplication sentences in their math journals.

7. In groups, students should continue to create 10 additional multiplication sentences.

8. Proceed to Activity 26 – Multiplication War.

MATERIALS

36 base-ten unit cubes per group

Number cubes (a pair per group)

Egg cartons (cut in half, showing 6 sections) 1 per pair

Math journal for each student

Pencils for each student

MATERIALS

Number cards
(Appendix G) or
decks of playing
cards (one set per
pair)

KEY WORD

Product

Multiplication War

1. Divide students into pairs. If a standard 52-card pack is used, take out the Kings, Queens, Jacks, and tens. Use the Ace as a one. Cards rank as usual from high to low: 9, 8, 7, 6, 5, 4, 3, 2 and Ace. Suits are ignored in this game.

2. Deal out all of the playing cards or number cards between players. Players do not look at their cards, but keep them in a pile facedown. The object of the game is to win all of the cards.

3. Both players turn their top card faceup and put them on the table. The player who says the product correctly first, gets to keep both cards.

4. If the turned up cards are equal, there is a war. The matching cards stay on the table and both players play the next card of their pile facedown and then another card faceup. The player who says the product of the new faceup cards quickest wins the war and adds all six cards facedown to the bottom of their pile. If the new faceup cards are equal as well, the war continues. Each player puts another card facedown and one faceup. The war goes on like this as long as the faceup cards continue to be equal. When the cards are different, the player who says the product first wins all of the cards in the war.

5. The game continues until one player has all the cards. This can take a long time. Give a time limit. Once students master this game go on to the extension activity below.

MATERIALS

4 copies of base-ten
cards (Appendix F)
per pair

Base-ten blocks

Extension

Use base-ten cards (Appendix F) to play Multiplication War. Add a new element to the game this time by putting the pile of base-ten blocks to the side. The student who says the product correctly first gets to take the corresponding base-ten blocks from the pile (i.e. 10 x 3 = 30. Winner takes three rods and zero units). Play continues until all of the cards or blocks have been used. The winner is the player who has the highest total of base-ten blocks.

Riddle Math

1. Divide students into pairs. Designate one student as Player A and one student as Player B. Distribute the corresponding Activity 27 Riddle Math page to each student. Have them cut apart the cards.

2. Distribute base-ten stamps or Appendix F to each student.

3. Players turn their backs to each other in order to hide their cards from one another and take time to try and solve their riddles independently. Each player should stamp or glue the answer to the back of the problem.

4. When problems have been solved independently, partners take turns presenting problems orally to see if the other layer can mentally compute the answer. If the player is able to do so, he or she gets to keep the card. The player with the most cards at the end of the game is the winner.

Name _____

Date _____

Riddle Math
Player A

My two dogs have 4 nails on each paw. How many nails do they have in all?	Kate and her friends went to the movies. They filled up 3 rows of 8 seats. How many people were in her group?
How many fingers does a family of 5 have total?	A recipe for one cheesecake requires 2 dozen eggs. How many eggs are needed to make one cheesecake?
Sam placed a total of 12 pickles on 4 hotdogs with an equal number of pickles on each hotdog. How many pickles were on each hotdog?	While getting his teeth cleaned, James counted 6 rows of 12 tiles on the ceiling. How many tiles were there in all?
At the circus, Kelly saw a total of 25 clowns exit 5 mini-cars. If they were grouped equally, how many clowns were in each car?	A total of 64 balloons were released every hour for an 8-hour time period. They were released in equal groups of how many balloons?
10 bees were flying into two honeycombs. The same number of bees went in each. How many bees were in each honeycomb?	14 socks were found on the floor. Once they were matched, how many pairs did they make?

Riddle Math
Player B

3 hungry spiders were busy making a web. How many legs do the spiders have in all? (Spiders have 8 legs.)	Jackson put 6 CD's each in 5 boxes. How many CD's does he have?
8 parrots laid 2 eggs each. How many eggs are there in all?	Tameka's bedroom has 72 books. Her mom asked her to stack them all evenly in her bookcase with 8 shelves. How many books will she place on each shelf?
36 marshmallows were put equally into 6 mugs of hot chocolate. How many marshmallows were in each mug?	Heldago spotted 3 starfish when he went snorkeling one day. Each starfish had 5 arms. How many arms did they have in all?
An adult human has 2 rows of 16 teeth. How many teeth does one adult have total?	On the first day of fall, 26 leaves fell from 4 different trees. The same amount fell from each tree. How many leaves fell from each tree?
Grandma is making a quilt. She made 70 squares. She is placing 7 squares in length. How many squares will she place in height?	A shark may have as many as 350 teeth at one time. If the shark has 7 equal rows of teeth. How many teeth are in each row?

Name _____

Date _____

Multiplication T-Tables

Find the pattern to help you fill in the T-Tables with the missing numbers.

1.

Rods	Units
1	10
2	20
3	30
4	
5	
6	
7	
8	
9	
10	
11	
12	

What is the pattern? _____

2.

Flats	Units
1	100
2	200
3	300
4	
5	
6	
7	
8	
9	
10	
11	
12	

What is the pattern? _____

3.

Tricycle	Wheels
1	3
2	6
3	9
4	
5	
6	
7	
8	
9	
10	
11	
12	

What is the pattern? _____

4.

Clocks	Numbers
1	12
2	24
3	36
4	
5	
6	
7	
8	
9	
10	
11	
12	

What is the pattern? _____

LESSON 2 Area

NCTM Standard: Area
Use visualization, spatial reasoning, and geometric modeling to solve problems.

GOAL
Students will use base-ten blocks to understand area.

DEVELOPMENTAL INSTRUCTION
Journal Question: How many units does it take to fill your ruler? How many rods does it take to fill your math journal? Do you know the math term for measuring the entire surface of an object?

1. Divide the class into groups of 3 or 4. Send the students on an "AREA HUNT."

2. Students should use flats, rods, and units to measure 5 objects around the room. Have students calculate the area of each object and record their findings in their math journals.

3. After 15 minutes, bring the groups back together.

4. Discuss the students' findings. Ask how the students found each area.

5. Instruct that the formula of the area of a rectangular shape is length multiplied by width, which gives the answer in square units or
A = l x w.

6. Hand out Activity 29 – Area Array.

7. Have students work independently, filling in the area of the multiplication problems with color pencils or crayons.

MATERIALS

Flats, rods, and units for each student

Math journal for each student

Pencil for each student

Ruler of each student

KEY WORDS

Area

Length

Width

Area Array

Use color pencils or crayons to show the area of each array described below. Fill in the missing product for each equation.

1. Use blue to create 12 x 6 = _____, red to create 3 x 4 = _____,

and purple to create 1 x 1 = _____.

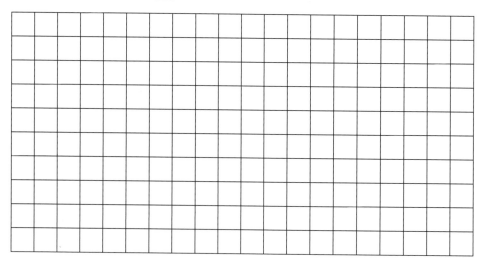

2. Use yellow to create 2 x 10 = _____, green to create 13 x 5 = _____,

and black to create 3 x 3 = _____.

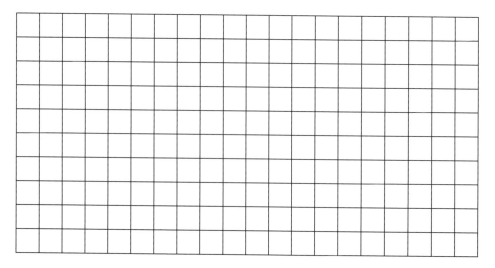

Area Practice

Use multiplication to find the area of the objects. To find the area of a rectangle you must multiply the length by the width. (A = 1 x w)

1. The length of the checkerboard is 8 cm. The height is also 8 cm. If you count the total number of squares, you will

    have _____. 8 x 8 = _____

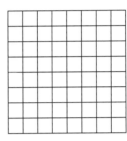

2. What is the area of a flat in square units? How did you solve it?

    _____

3. What is the area of a rod in square units? How did you solve it?

    _____

4. What is the area of this flag and its pole in square units?

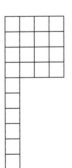

    _____

5. What is the area of this patchwork quilt in square units?

    _____

Name _____

Date _____

Area Animals

Use flats, rods, and units to create an animal below.
Draw the animal and record its area.

EXAMPLE: Teddy Bear = 35 square units

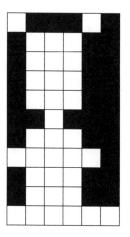

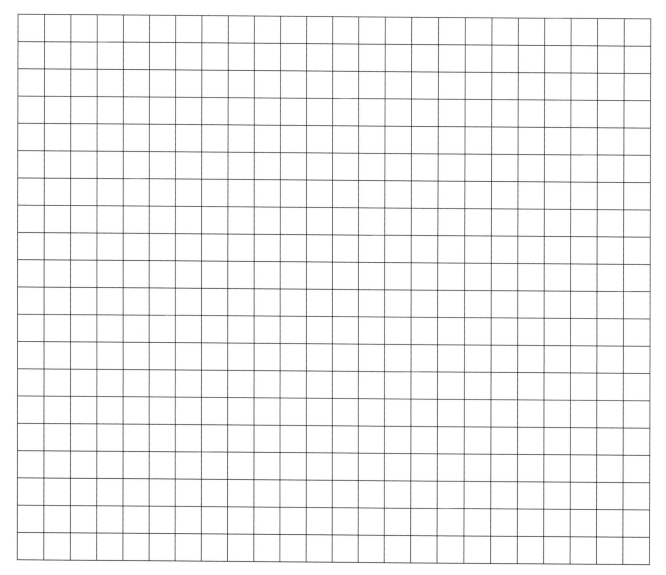

LESSON 3 Multiplication With Regrouping

NCTM Standard: Multiplication
Develop fluency in multiplying whole numbers.

GOAL
Students will practice multiplying two- and three-digit numbers.

DEVELOPMENTAL INSTRUCTION

1. Give base-ten manipulatives to each student (1 flat, 10 rods, and 15 units).

2. Place the same number of base-ten blocks on the overhead projector.

3. Instruct students that these tools can be used to find the product of two factors.

4. Demonstrate by using the problem: 15 x 3 = 45. Follow the visual model shown. Students should mimic the steps using the base-ten materials.

5. The number 15 can be broken into 10 and 5. Place three rods and three groups of 5 units to show 3 x 10 plus 3 x 5.

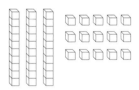

6. Regroup the fifteen ones to make another rod and five ones.

7. Add the one ten that was regrouped for 4 rods and 5 units or 45.

8. Pair students together, combining base-ten blocks. Continue modeling new problems, such as 12 x 4, 53 x 2, 32 x 4, 22 x 6, etc.

9. Proceed to Activity 32 – Multiplication Workout.

MATERIALS

Overhead projector

Base-ten blocks

Activity 32 worksheet

KEY WORD

Regroup

Multiplication Workout

Solve each problem.

1.	24	2.	19	3.	25	4.	16
	x 4		x 3		x 6		x 2

5.	36	6.	29	7.	91	8.	13
	x 8		x 7		x 5		x 9

9.	35	10.	18	11.	39	12.	48
	x 4		x 6		x 3		x 5

13.	38	14.	62	15.	45	16.	54
	x 2		x 4		x 3		x 5

17. If one carton holds 22 strawberries, how many strawberries would be in 6 cartons?

18. If one candy bar costs $.65, how much would it cost you to buy one for each member of your family?

LESSON 4 Division

NCTM Standard; Division
Develop fluency in dividing numbers.

MATERIALS

Base-ten blocks or Appendix F

Activity 33 worksheet

GOAL
Students will practice dividing two-digit numbers by one-digit numbers.

DEVELOPMENTAL INSTRUCTION

1. Distribute base-ten materials to students.

2. Tell students, "Division is sharing equally. We will use the base-ten blocks to represent the numbers in the following story problems. Follow along as we solve these problems together."

KEY WORDS

Division

Regroup

Example: "There are 9 bones and 3 dogs. How many bones are there for each dog?"

Example: "8 cookies were shared equally by 4 people. How many cookies did each person get?"

3. After modeling and discussing the problems, present more difficult story problems. Inform students that for the next two problems they will have to regroup.

Journal Question: Using base-ten blocks, how would you show regrouping in the next two problems?

a. 27 flowers were planted equally among three planters. How many flowers are in each planter?

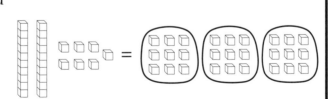

b. Share 36 cashews equally among 6 hungry adults. How many cashews will each person get?

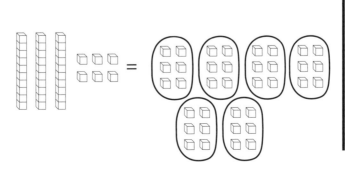

Two-Digit by One-Digit Division

Use base-ten blocks to help you divide these problems.

Look at the example and follow the steps for each person.

EXAMPLE:

Tens | Ones
        1 | 2
3 | 36

Step 1: Start with 3 rods and 6 units. Divide the tens.
Place 1 rod or ten in each group of 3:

Step 2: Divide the ones. Place 2 units or ones in each group:

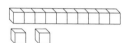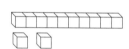

1. 2⟌40

2. 2⟌20

3. 5⟌55

4. 3⟌66

5. 4⟌52

6. 3⟌35

Division T-Tables

Find the pattern to help you fill in the T-Tables with the missing numbers.

1.

Units	Rods
100	10
200	20
300	30
400	
500	
600	
700	
800	
900	
1,000	
1,100	
1,200	

What is the pattern? _____

2.

Units	Flats
100	1
200	2
300	3
400	
500	
600	
700	
800	
900	
1,000	
1,100	
1,200	

What is the pattern? _____

3.

Wheels	Cars
4	1
8	2
12	3
16	
20	
24	
28	
32	
36	
40	
44	
48	

What is the pattern? _____

4.

Pages	Books
25	1
50	2
75	3
100	
125	
150	
175	
200	
225	
250	
275	
300	

What is the pattern? _____

MATERIALS

Base-ten blocks or
Appendix F

Transparency of
Appendix M

Overhead projector

KEY WORDS

Remainder

Leftover

LESSON 5 Division with Remainders

NCTM Standard: Division
Develop fluency in dividing numbers.

GOAL

Students will practice dividing two- and three-digit numbers by one-digit
numbers with remainders.

DEVELOPMENTAL INSTRUCTION

1. Select 5 students to stand in the front of the classroom.

2. Say, "I need to divide these students into two groups. How many
   students will be in each group? How many students will be leftover?"

3. Have a volunteer from the class divide the group into two equal parts
   and explain how many remain.

4. Define *remainder in division* as the leftover amount after dividing
   equally.

5. Have students use base-ten blocks to solve division problems with
   regrouping and remainders. Model the problem 152/3 on the overhead
   projector. Use base-ten blocks to demonstrate the following steps:

   • BUILD THE NUMBER: 1 flat in the hundreds column, 5 rods in the
     tens column, and 2 units in the ones column.

   • REGROUP: Remove the one flat from the hundreds and replace it
     with 10 rods. Place the rods in the tens columns.

   • SOLVE: Circle even groups of rods to solve the problem. Discuss
     the remainder. Students should circle 3 groups of 5 rods each.

6. Model 379/3 with an overhead transparency of the Base-Ten Place-
   Value Mat (Appendix M) and base-ten blocks. Discuss the remainder.

7. Practice with Activity 35 – Tic-Tac-Toe Division with Remainders.

Names _____

_____

Date _____

Tic-Tac-Toe Division with Remainders

Work independently to solve the division problems below. You may use base-ten blocks to help you.

5$\overline{)474}$	2$\overline{)789}$	3$\overline{)374}$
6$\overline{)343}$	8$\overline{)659}$	9$\overline{)123}$
3$\overline{)134}$	4$\overline{)857}$	5$\overline{)321}$

Now, find a partner and compare your answers. Make two copies of the pieces below to play a game of tic-tac-toe. Player A colors their pieces red and player B colors their pieces blue. Player A begins by placing his or colored answer on the correct problem. Play alternates between players until one covers three spaces in a row.

214 R1	13 R6	124 R2	394 R1

57 R1	64 R1	82 R3	94 R4	44 R2

Name _____

Date _____

Chapter 4 Assessment:
Multiplication and Division Number Tiles

1.   39
        x 3

2.   3⟌17

3.   65
        x 3

4.   5⟌35

5.   28
        x 4

6.   4⟌12

7.   41
        x 9

8.   9⟌39

Use number tiles from Appendix V to complete problem A. Write the numbers in the squares. Then use the same tiles to record the correct inverse solution in problem B. Not all the squares have to be filled in problem B.

A.

B.

CHAPTER 5

Review Games
and
Weekly Workouts

■ Review Game Directions

Game pieces for Tour of England and Race to the Finish Line

Game Markers – A   Spinner   Game Markers – B

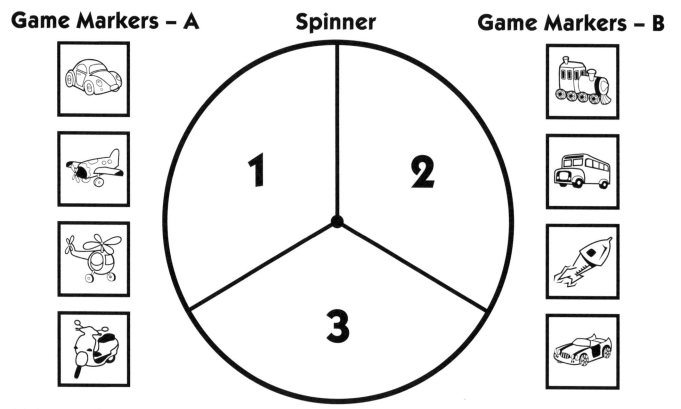

MATERIALS

Review Game Board
Game Markers or unit cubes
Review Game Cards
Spinner

One calculator per student
Pencils
Paper
Paperclip to use with spinner

DIRECTIONS

- 2-4 students may play at a time.

- Choose game board A or B. Use unit cubes or the game pieces above as markers for each player.

- Copy the game cards and cut them apart.

- Players take turns presenting the game cards to each other and solving the math problems. Remaining players check answers using calculators, until it is their turn. The answer is printed on the card.

- If the player answers correctly, he or she may spin the spinner and advance the amount shown on the spinner.

- If the problem is answered incorrectly, play goes to the next student.

- The winner is the player who reaches either the castle or the finish line first.

■ Review Game Board A

Tour of England

You have reached the castle! You are a winner!

Meet the Queen for tea. Go ahead two spaces.

Share a fish and chips dinner with a friend. Go ahead one.

Fire breathing dragon! Go back one space.

Give royalty a ride to a cricket match. Lose a turn.

Stop to make a phone call. Lose a turn.

Your English Tour Begins Here.

Start

■ Review Game Board B

Race to the Finish Line!

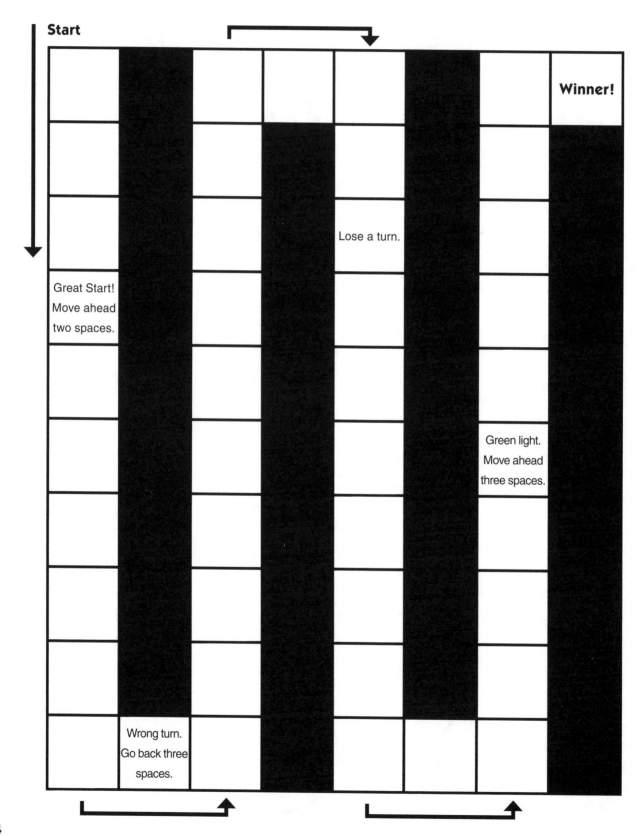

Start

Winner!

Lose a turn.

Great Start! Move ahead two spaces.

Green light. Move ahead three spaces.

Wrong turn. Go back three spaces.

Review Game Cards

How much money do you have if you have six quarters and two dimes? *($1.70)*	Two flats, three rods, six units + nine flats, four rods, one unit = ? *(1,177)*
How much money do you have if you have ten pennies and three dimes? *($.40)*	Four flats, two rods, nine units + One flat, six rods, two units = ? *(591)*
How much money do you have if you have two dollar bills, ten dimes and ten pennies? *($3.10)*	Seven flats, five rods, four units – Three flats, two rods, three units = ? *(431)*
$2.55 + $6.82 = ? *($9.37)*	Nine flats, nine rods, zero units – One flat, one rod, nine units = ? *(871)*
71.4 + 2.11 = ? *(73.51)*	Six flats, four rods, two units – Four flats, zero rods, six units = ? *(236)*

234 x 4 = ? *(936)*	789 ÷ 3 = ? *(263)*
581 x 2 = ? *(1,162)*	Make this number using base-ten blocks 458 *(four flats, five rods, eight units)*
236 x 5 = ? *(1,180)*	Make this number using base-ten blocks 699 *(six flats, nine rods, nine units)*
228 ÷ 4 = ? *(57)*	Make this number using base-ten blocks 481 *(four flats, eight rods, one unit)*
561 ÷ 3 = ? *(187)*	Make this number using base-ten blocks 107 *(one flat, zero rods, seven units)*

■ Review Game Cards

What number is represented by three flats, two rods and four units?
(324)

What is the perimeter of a rectangle that has a length of 3 cm and a width of 2 cm?
(10 cm)

What number is represented by six flats, zero rods and one unit?
(601)

What is the area of a square with a length of 10 cm?
(100 square cm)

What number is represented by eight flats, eleven rods and zero units?
(910)

What is the area of a rectangle with a length of 4 cm and a width of 6 cm?
(24 square cm)

Which number is greater: two flats, twenty rods, and ten units (410) or three flats, six rods and four units (364)?
(410)

If you had $3.50 and you bought a toy car for $.75, how much money would you have left?
($2.75)

What is the perimeter of a square that has a length of 4 cm?
(16 cm)

If you had $10.00 and you bought a hamburger for $4.00, a soda for $1.00 and fries for $.50, how much money would you have left?
($4.50)

WEEKLY WORKOUT 1
Chapter 1

Name _____

Date _____

Day 1

1. A 5 in the ones place has a value of _____ .
2. A 3 in the tens place has a value of _____ .
3. A 7 in the hundreds place has a value of _____ .
4. A 1 in the thousands place has a value of _____ .
5. When you add each of these numbers together, your total is _____ .

Day 2

Record the number represented by each of the base-ten blocks.

1.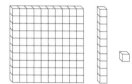

2.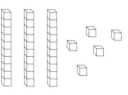

3.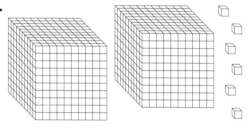

_____

_____

_____

4.

5.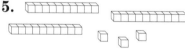

_____

_____

Day 3

1. $300 + 20 + 5 =$ _____
2. $1,000 + 800 + 70 + 2 =$ _____
3. $2,000 + 10 + 4 =$ _____
4. $7,000 + 100 + 3 =$ _____
5. $5,000 + 600 + 20 + 1 =$ _____

Day 4

Complete the patterns.

1. 20; 30; 40; _____, _____, _____, _____
2. 210; 220; 230; _____, _____, _____, _____
3. 197; 198; 199; _____, _____, _____, _____
4. 405; 430; 455; _____, _____, _____, _____
5. 1,000; 1,100; 1,200; _____, _____, _____, _____

Day 5

1. Which digit is in the hundreds place? **297** _____
2. Place the numbers in order from least to greatest.
   a. 82; 272; 22; 2,007; 128 _____
   b. 310; 103; 301; 110; 3 _____
   c. 16; 1,006; 601; 116; 61 _____
3. Write this number in words. **524** _____

Name _____

Date _____

Day 1

1. A 7 in the ones place has a value of _____.
2. A 2 in the tens place has a value of _____.
3. A 0 in the hundreds place has a value of _____.
4. An 8 in the thousands place has a value of _____.
5. When you add each of these numbers together, your total is _____.

Day 2

Record the number represented by each of the base-ten blocks.

1. 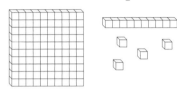 **2.** 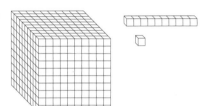 **3.**

_____   _____   _____

4. 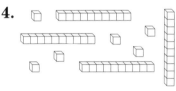 **5.**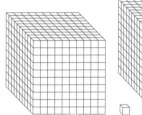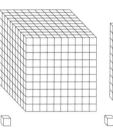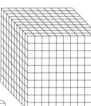

_____   _____

Day 3

1. 8,000 + 300 + 50 + 4 = _____   2. 500 + 60 + 1 = _____
3. 6,000 + 400 + 40 + 2 = _____   4. 1,000 + 8 = _____
5. 2,000 + 300 + 20 + 1 = _____

Day 4

Complete the patterns.

1. 35; 45; 55;   _____, _____, _____, _____
2. 611; 612; 613;   _____, _____, _____, _____
3. 1,097; 1,098; 1,099; _____, _____, _____, _____
4. 275; 280; 285;   _____, _____, _____, _____
5. 942; 944; 946;   _____, _____, _____, _____

Day 5

1. Which digit is in the tens place? **1,304** _____
2. Write > or < in each space.

   a. 23 ◯ 32   b. 1,002 ◯ 1,001   c. 987 ◯ 978   d. 213 ◯ 302

Day 1

1. Which number shows hundredths? _____
2. Which number shows ones? _____
3. Which number shows tenths? _____
4. Which number shows hundreds? _____

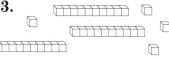

1,993.061

Day 2

Record the number represented by each of the base-ten blocks.

1.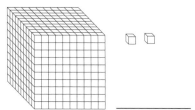

_____

2.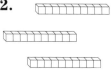

_____

3.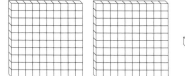

_____

4.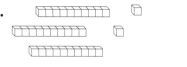

_____

5.

_____

Day 3

Write the decimal amount for each set.

1.

_____

2.

_____

3.

_____

4.

_____

Day 4

Write the equivalent decimal.

1. $\dfrac{5}{100}$ 2. $\dfrac{6}{10}$ 3. $\dfrac{25}{1,000}$ 4. $\dfrac{9}{10}$ 5. $\dfrac{57}{100}$

_____ _____ _____ _____ _____

Day 5

Write > or < in each space.

1. 6.78 ◯ 67.8   2. 43.1 ◯ 14.3

3. 99.8 ◯ 98.8   4. $21.04 ◯ $21.40

5. 14.03 ◯ 1.403

80

WEEKLY WORKOUT 2
Chapter 2

Name _____

Date _____

Day 1

1. Which number shows hundreds? _____
2. Which number shows ones? _____
3. Which number shows tenths? _____
4. Which number shows tens? _____

$$\boxed{623.17}$$

Day 2

Record the number represented by each of the base-ten blocks.

1. _____

2. _____

3. 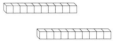 _____

4.  _____

5. 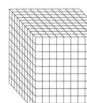 _____

Day 3

Write or draw another way to show each amount.

1. _____

2. _____

3. 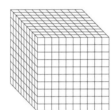 _____

4. _____

5.   _____

Day 4

Write the equivalent decimal.

1. $\dfrac{15}{100}$ 2. $\dfrac{7}{10}$ 3. $\dfrac{48}{1{,}000}$ 4. $\dfrac{2}{10}$ 5. $\dfrac{936}{1{,}000}$

_____   _____   _____   _____   _____

Day 5

Write > or < in each space.

1. 1.23 ◯ 12.3   2. 32.6 ◯ 3.266

3. 334.2 ◯ 33.42   4. $2.39 ◯ $2.93

5. 2.99 ◯ 1.99

Day 1

Round to the nearest ten.

1. 24 _____   **2.** 25 _____   **3.** 48 _____

Round to the nearest hundred.

4. 354 _____   **5.** 247 _____   **6.** 671 _____

Day 2

Solve.

1. 121	**2.** 345	**3.** 672	**4.** 590	**5.** 786
+ 78	+135	+ 29	+337	+ 54

Day 3

Estimate the sum by selecting the most likely choice.

1. 578 + 542	**2.** 45 + 24 + 33	**3.** 121 + 246 + 275	**4.** 123 + 456	**5.** 463 + 116
a. 1,000	a. 90	a. 600	a. 400	a. 500
b. 1,200	b. 110	b. 500	b. 500	b. 600
c. 1,100	c. 120	c. 700	c. 600	c. 400
d. 900	d. 100	d. 400	d. 700	d. 700

Day 4

Solve.

1. $4.35	**2.** $2.76	**3.** $4.09	**4.** $.76	**5.** $12.33
+$.45	+$4.33	+$6.11	+$.65	+$ 3.92

Day 5

Write > or < in each space.

1. 23.32 + 4.29 ◯ 23.23 + 4.92   **2.** 3.14 + 6.55 ◯ 3.41 + 65.5

3. .40 + .60 ◯ .33 + .74   **4.** 3.49 + 6.21 ◯ 3.55 + 5.99

5. 3.72 + 4.53 ◯ 4.1 + 3.78

Day 1

Round to the nearest hundred.

1. 657 _____ **2.** 328 _____ **3.** 461 _____

Round to the nearest thousand.

4. 2,594 _____ **5.** 1,352 _____ **6.** 5,618 _____

Day 2

Solve.

1. 365	**2.** 846	**3.** 331	**4.** 894	**5.** 990
+124	+254	+225	+362	+119

Day 3

Estimate the sum by selecting the most likely choice.

1. 145 + 789	**2.** 474 + 98	**3.** 515 + 151 + 949	**4.** 346 + 183	**5.** 27 + 14 + 82
a. 1,000	a. 600	a. 1,000	a. 400	a. 100
b. 900	b. 700	b. 1,400	b. 500	b. 110
c. 800	c. 500	c. 1,500	c. 600	c. 120
d. 2,000	d. 400	d. 1,600	d. 700	d. 130

Day 4

Solve.

1. $6.28	**2.** $8.31	**3.** $5.17	**4.** $.99	**5.** $24.45
+ $ .62	+ $3.67	+$4.38	+$.99	+$12.66

Day 5

Write >, <, or = in each space.

1. 68 + 2.54 ◯ 70.5 + .87 **2.** 19.1 + 42.88 ◯ 18.9 + 43.2

3. .55 + .75 ◯ .6 + .8 **4.** 3.26 + 1.24 ◯ 4.5

5. 8.48 + 24.2 ◯ 25.6 + 3.16

Name _____

Date _____

Day 1

1. Another way to solve 11 + 11 + 11 + 11 = 44 is _____.

2. Another way to solve 12 + 12 + 12 + 12 + 12 = 60 is _____.

Show using repeated addition.

3. 4 x 20 = _____

4. 25 x 3 = _____

Day 2

Write a multiplication problem for each of the pictures.

1.   2. 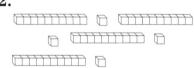   3.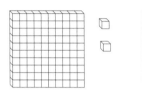

_____   _____

_____

Day 3

Complete the T-tables.

Hands	Fingers
1	5
2	10
3	
4	
5	

Decks	Cards
1	52
2	104
3	
4	
5	

Day 4

How much did you spend?   **Cola = $.25   Milk = $.30   Juice = $.45**

1. Five colas _____   2. Seven milks _____   3. Two juices _____

4. One cola and three milks _____   5. Four juices and two colas _____

Day 5

Find the area and write it inside the rectangle.

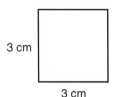

3 cm
3 cm

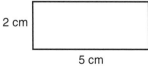

2 cm
5 cm

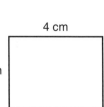

4 cm
3 cm

6 cm
8 cm

Note: Figures are not drawn to scale.

84

Name _____

Date _____

Day 1

1. Another way to solve $100 - 20 - 20 - 20 - 20 - 20 = 0$ is _____ .

2. Another way to solve $48 - 6 - 6 - 6 - 6 - 6 - 6 - 6 - 6 = 0$ is _____ .

3. Another way to solve $50 - 10 - 10 - 10 - 10 - 10 = 0$ is _____ .

Show using repeated subtraction.

4. $45 \div 5 =$ _____

5. $375 \div 125 =$ _____

Day 2

Write two story problems using the number of students in your class. Use multiplication in the first problem, and division in the second problem.

1. _____

   _____

2. _____

   _____

Day 3

Complete the T-tables.

Cookies	Dozen
24	2
36	
48	
60	
72	

Crayons	Boxes
64	1
128	2
192	
256	
320	

Day 4

How much did you spend? **Cookies = \$.15   Cake = \$.75   Candy = \$.65**

1. Nine cookies _____

2. Three cakes _____

3. Five candies _____

4. Three candies and one cookie _____

5. Four cakes and two candies _____

Day 5

Find the area and write it inside each rectangle.

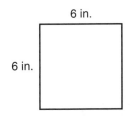
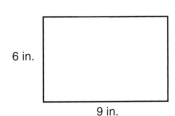
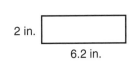

Note: Figures are not drawn to scale.

340	**582**	**104**
97	**1,406**	**12**
955	**75**	**210**

300 + 40	500 + 80 + 2	100 + 4
90 + 7	1,000 + 400 + 6	10 + 2
900 + 50 + 5	70 + 5	200 + 10

Base-Ten Rummy

Activity Mat

Hundreds	Tens	Ones

Hundreds	Tens	Ones

Hundreds	Tens	Ones

Hundreds	Tens	Ones

Hundreds	Tens	Ones

Hundreds	Tens	Ones

Hundreds	Tens	Ones

Hundreds	Tens	Ones

Hundreds	Tens	Ones

Hundreds	Tens	Ones

Hundreds	Tens	Ones

Hundreds	Tens	Ones

Hundreds	Tens	Ones

Hundreds	Tens	Ones

Hundreds	Tens	Ones

Hundreds	Tens	Ones

Hundreds	Tens	Ones

Hundreds	Tens	Ones

Base-Ten Cards

Number Cards

0	**0**	**1**	**1**
2	**2**	**3**	**3**
4	**4**	**5**	**5**
6	**6**	**7**	**7**
8	**8**	**9**	**9**

Greater Than   **Less Than**

Dollar Puzzle

Money Cards

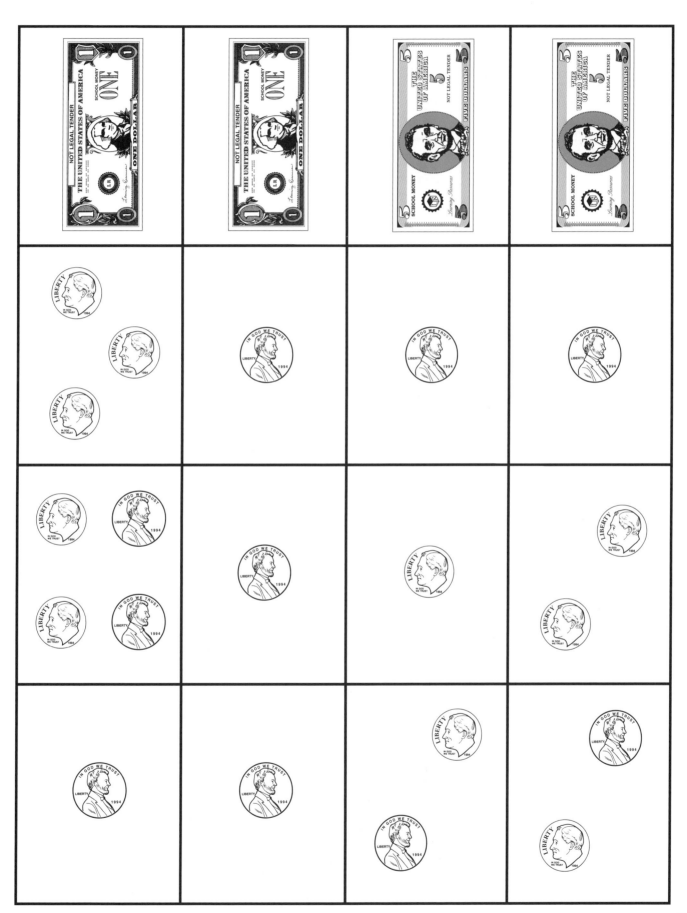

.50	$\dfrac{1}{2}$	50%	
.75	$\dfrac{3}{4}$	75%	
1.00	$\dfrac{100}{100}$	100%	
.10	$\dfrac{1}{10}$	10%	

Rounding Hill

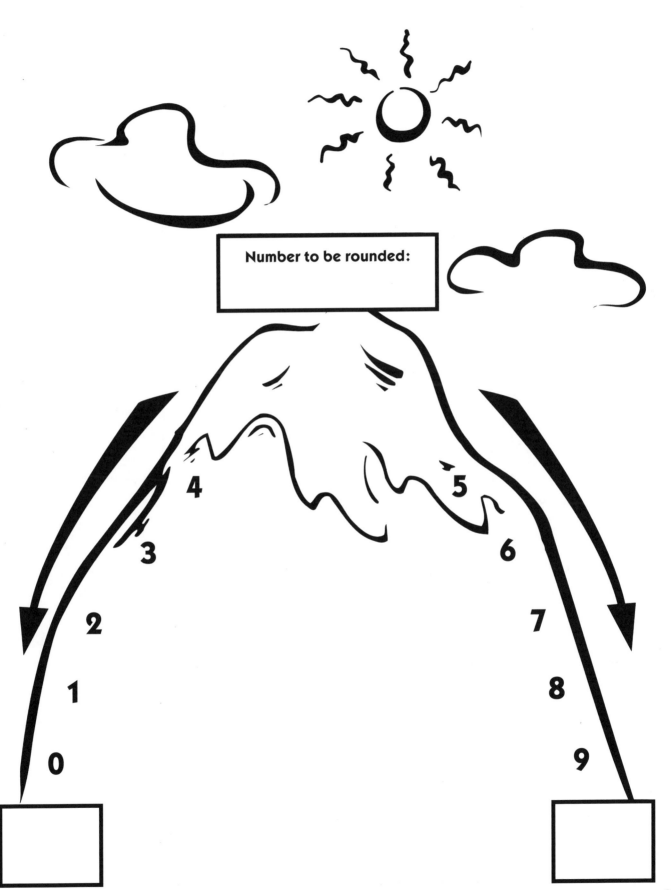

Number to be rounded:

0 1 2 3 4 5 6 7 8 9

Base-Ten Place-Value Mat

Cubes (Thousands)	Flats (Hundreds)	Rods (Tens)	Units (Ones)

Estimating Store

Climb the Mountain

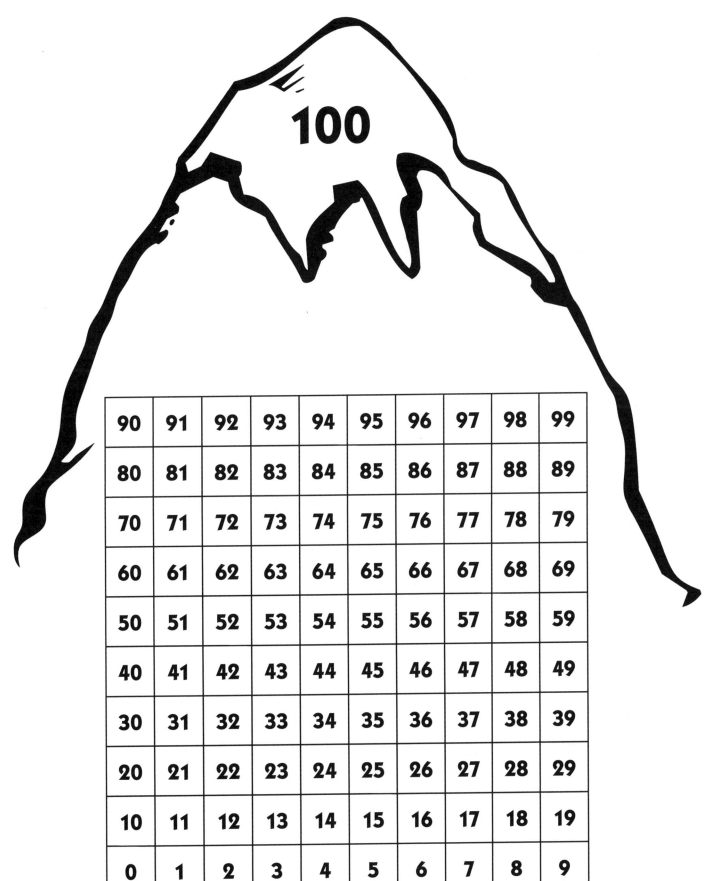

90	91	92	93	94	95	96	97	98	99
80	81	82	83	84	85	86	87	88	89
70	71	72	73	74	75	76	77	78	79
60	61	62	63	64	65	66	67	68	69
50	51	52	53	54	55	56	57	58	59
40	41	42	43	44	45	46	47	48	49
30	31	32	33	34	35	36	37	38	39
20	21	22	23	24	25	26	27	28	29
10	11	12	13	14	15	16	17	18	19
0	1	2	3	4	5	6	7	8	9

Meet the Neighbors

235	**195**	**408**
302	**111**	**625**
32	**56**	**17**
29	**64**	**29**

51	**12**	**22**	**34**
7	**40**	**100**	**0**
36	**3**	**48**	**24**
5	**11**	**9**	**42**
10	**8**	**15**	**60**

Centimeter Grid Paper

Appendix T

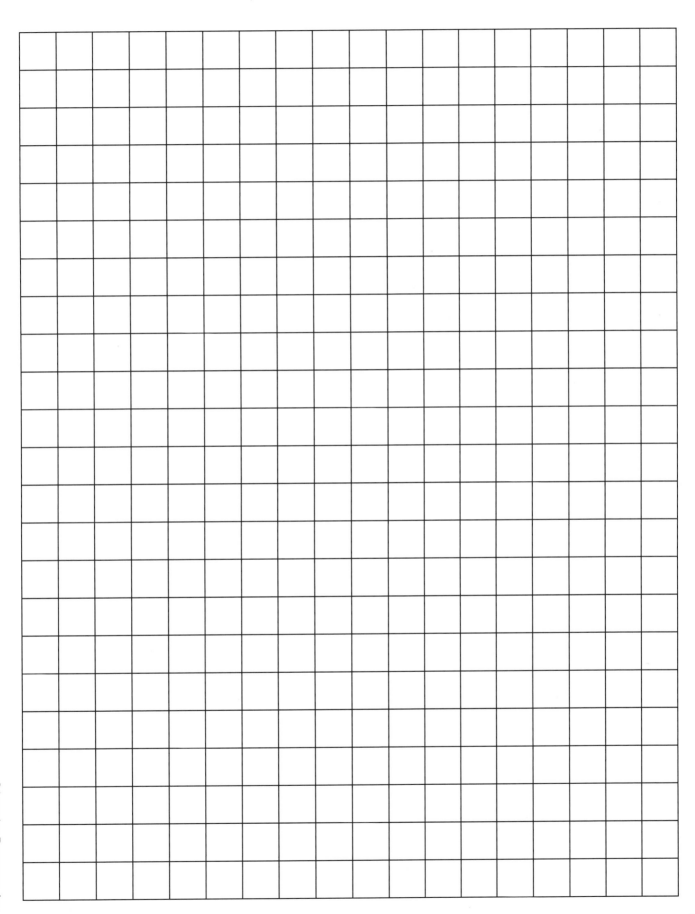

Spinners

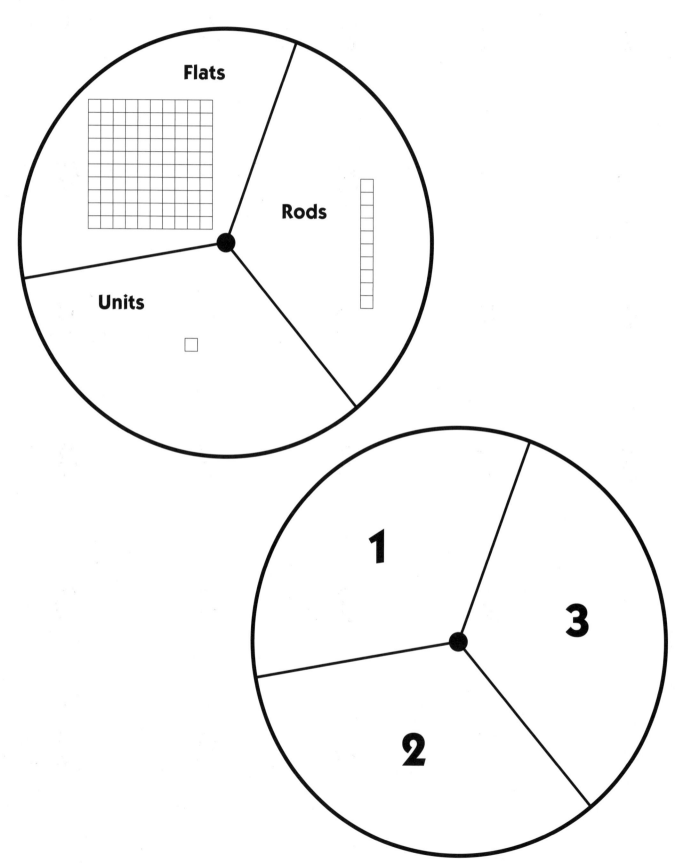